The Nemo Poems:
A Martian Perspective

The Nemo Poems: A Martian Perspective
by Rodger Martin

ISBN: 978-1-960293-16-9
Second Edition

Published by NatureCulture® LLC
NatureCulture Web Imprint
www.nature-culture.net

Cover Design: Christopher Gendron
Cover Imagery: NASA:
Earth: NASA/JPL; Mars: NASA/JPL-Caltech/University of Arizona

Interior book design: Lis McLoughlin

The Nemo Poems:
A Martian Perspective
by
Rodger Martin

NatureCulture
Northfield, Massachusetts

Author's Foreword to the Second Edition

We like to think we have come vast intellectual and emotional distances from our ancestors, and yet, when the face of death confronts us, we feel—if not see—how little we have changed from those ancestors and how tiny a piece of the universe within which we actually exist. It is at that moment when poetry and story become the umbilical making survival worthy of the effort to survive.

And so it is with Nemo, a boy turned soldier searching the planet in hopes of understanding his childhood losses and discovers only more loss until he unexpectedly senses another, even more ancient stranger who has fled his own world of war and loss. They both begin to understand that if they are to heal, the old aphorism "Home is where the heart is" must become a two-way street which embraces "Heart is where the home is."

The ancient, elder veteran realizes that only by serving as a guardian of his younger veteran's emotions, they both might keep their hearts and eventually find their way home—that place where like Tolstoy's Pierre Bezukhov and Homer's Odysseus—they might finally bury their oars.

—Rodger Martin
2/25/25

Publisher's Foreword to the Second Edition

NatureCulture® chose to publish this book under our NatureCulture Web imprint as part of a commitment to help people be in right relationship with Nature.

War and climate change are twin existential threats that are human-caused, and as such are preventable. We contend that both may be more effectively addressed by humans in a way that acknowledges that we are part of natural life on Earth, and that when we act like we are not, the world we create is damaged and out of balance.

It should not take an "outsider" perspective to show us how we have not tended to our planet. But in case it would help, please read this book.

—*Lis McLoughlin, PhD, Founder and Director*
NatureCulture® and NatureCuture Web
2/25/25

Sing to me of the man, Muse, the man of twists and turns
driven time and again off course, once he had plundered
the hallowed heights of Troy.
Many cities of men he saw and learned their minds,
many pains he suffered, heartsick on the open sea,
fighting to save his life and bring his comrades home.

—The opening lines of Homer's *The Odyssey*
Translated by Robert Fagles
Penguin Classics Deluxe Edition, NY: 1996, p.77.

The Nemo Poems: A Martian Perspective
Table of Contents

I.
Calling

The Meeting

My name like your Sistine face,
is a fingertip touch that warps time
and soars like Icarus. But I escape

thin atmosphere to indigo space
which ices all but feeling and night
in this alien search for home.

Out here loneliness is a giant wheeling,
an asteroid purged of flame and place.
Here twisted *Voyager* nosed frozen Titan,

a welfare mother picking the Cobras'
cosmic garbage searching for a sealing
to bring closure and tilt the crown of Saturn.

I first caught your iron, later, within lighter metal
I disrobed your secrets, caressed your stomach
and found you sent by a softness for twos.

Now, I drift forth again from the red rift
Valles Marineris, from vapors trailing
Olympus Mons beneath an orange sky,

to seek this journal of Diana and Nemo
understand their simple solace, their why
and for both a tracking backward in time.

The Landing
Hancock Radio Telescope

The salamanders' bulbous eyes gleaned
        the flash of methane here, little removed
        from the time their ancestors, like mine,
        ionized from primal mist, slithered
        through dim forests of fern, leaving
        mucous glistening on their starlit trails.

Now, the environmental wars have exploded again.
        and this place has turned firebase ringed with wire.
        Flame keeps out the night; a dungeon bunker
        lies beneath pocked mud and mows down
        wood until it falls away from the mire.
        A dozer's steel splinters every form
        that tries to stand forcing undines
        to dig in and spy from beneath the leaves.

An electrical hum reminds the jungle of what
        it's lost. Darkness darkens again. A giant
        metal hand reaches from the concrete,
        cups a skeletal eye, stretches toward the stars.
        It opens to light so dim, so old it might blink
        for amphibious time, might wobble for a pulsing
        Chloroplast, but will hardly wink at bipeds,
        weapons slung between legs, who scar
        the loam and scorch their village to save it.

The salamanders creep out and up the bunker walls.
        Bipeds lock and load; they strain over screens,
        squint through slits into the black. Listen
        for slither and hiss of lidless eyes and miss
        the slime that moistens the back of their dreams.

I have seen it all before. This time Nemo must sleep and go back alone,
feel the chill himself.

Football Practice

Sleet whistles across the frozen flats.
The isolated players sit in silent formation.
    Their splint-white legs stretched fore.
    Surviving soldiers
On the northern steppes of a lost war.
Their coaches, Pattoned against the gusts,
    Front these sprouts. Search
    For that—last ounce—
    That—final move.

Strange battle these wait to fight
Coming home not dead but beaten.
    Their women wait;
    Their letters—wait;
    Their—fame—waits,
    While they splay,

stacked shells cut against an Arctic slate
forcing their bitten stumps to drive
at that elusive ghost in the dark
clouds sprinting above the plain.

In The Guise of The Canine

I own you well, wolf,
nuzzled in these dreams
and the leafless stem of time.

In the 3 a.m. dark
my pad becomes your tread.
your smooth, worn claws
glisten in the auroral light.

From Saginaw to Denali
your night wail echoes
off the canyon wall.

I watch, through your dark cornea,
the elk pick in the mist-choked swamp,

      And late at moon, wolf,

when the silence of my kind
erases the present, I taste
from your tongue
and feel the incisor cut
living from the dead.

Lyme Regis,
before Nemo

The moon dispensed blessings to each tufted wavelet
that channeled to worship the stone cobb.
All about, the tidal sanctuary glistened,
buttressed against the striated shale cliffs that fanned
beneath the studded, crystal skies. Diana rode
her mare along the crest and pondered the beachside lights
of a town like her which survived kings and wars and sea.

Just woman, Saxon blood, her blonde genes
boiled with dreams of foal, and hare, and lithe men
whose buttocks rippled. She had yet to foresee the night
young Clark stumbled from the Rock Point Inn
and she hid in the Norman shadows saying, "He's drunk
again; don't let him see me." But this night
only shells echoed, and she nudged her mount toward town.

Nemo's dinner in a Brattleboro Diner
with Diana, daughter of a Nazi engineer

Our mothers lie across an Atlantic green, cold, and deep,
A canyon half the distance grass and petal keep
When we tend their simple lots. Your father chose SS.
His boot divided steppes. Mine picked Eighth Air Force.
His belly burned babies, turned your mother to ash.
Mine survived the Blitz, but that dust, your father's tread
From a thousand asphalt dawns, that cancerous wash,
Scoured out her breasts. They meet before us now spread
On a bridge built in London, Stalingrad, Dresden.
As a refugee, you planted saplings, defied your father,
Dug potatoes until you dropped. My father's axe:
"It is you children who break your mothers' backs."

I am mirror above the table. Children do not set the stage.
Your fathers play at war. Your mothers are the wage.

II.
In-Country

At the Laterite Pit*

Sweat-worn soldiers lounged both sides of the road. A water buffalo turned the bend, its bell chimes soft as breeze. It bobbed between the fatigues and pulled a two-wheeled cart. Atop sat a peasant woman, old, bouncing with her load. The day steamed hot and she sensed an angry heart and urged her buffalo into trot.

One boy rose, scooped a baseball-sized clot, "Hey bitch," he jeered. "Run, hell, d'ya forget your keys?" He carefully aimed.

For a moment this place ended and a ball field returned, groomed as the states. A strap of a pitcher, jersey streaked with dirt from a slide in the second base dust; two strikes already rung at the plate, he kicked high with his brother's worn spikes.

But memory seasons and every throw ends with a pitch. It struck the beast square on its flank. The buffalo kicked at the pain, lunged toward the bank. Amid licensed laughter, woman and cart, careened in the ditch.

I must look further.

Laterite was the name given to a red-colored earth in Vietnam, primarily used for construction purposes.

Barbed Wire

In the drunken night brawl of battle,
no place to drop but onto the concertina.
He flattened himself against the barbs
And even in the dark, his reflexes worked,
Caught the glint of one glinting needle and shifted.
It cut just under his eye. Missed his vision
And in the sudden silence when soldiers sense
There is nothing left to shoot, opened a vision past.

     He was six, barefoot, at run in the Pennsylvania verdancy:
     church, bread, limestone-sweetened corn, Jerseys
     spacing the pastures. Awry of judgment he jogged,
     did not see the wire strand, its points sharp
     like Gabriel's spear barring return to innocence.
     It struck just under his eye. He learned then
     To trust a reflex. The pain shocked less than blood,
     Both less than the assault—that humanity-inspired
     Cold, metallic, mathematical dagger hanging in the air,
     Portent of the humanity to come stacking bodies
     Soaked like wet cigars against the wire of its name.

Prek Klok*

I wonder whether the airplane or the rifle is the biggest dick in this war thought Nemo as he stepped onto the baked airstrip and wove among the squatting klatches of irregular Vietnamese soldiers and their families. He blocked the high pitch of a language that to him was always jabber and walked up the dropped, rear hatch of an olive drab twin-engined army Caribou transport. The nineteen-year-old scanned the interior of the plane for a place to sit. In the center his squad sergeant bent over and checked the thin, metal bands which fastened the bucket loader to the floor. In front of the loader two stacks of eight-foot-long, three-foot-wide sheets of psp—pierced steel planking—were also banded onto the floor of the cargo bay. Like thin green slices of Swiss cheese, the plates had lines of round holes punched out of them to reduce weight. They then could be strung together over raw earth to form a quick landing strip for planes like this Caribou.

Nemo cursed. He'd spent an hour breaking his back loading the pieces, and in a few minutes he'd break his back again unloading them on the airstrip at Prek Klok. He knew little about Prek Klok except it was a Special Forces camp and the VC had chewed up the airstrip. His sergeant volunteered the squad to fix it.

Nemo looked toward a seat against the front of the compartment, facing the banded psp, beneath the cockpit and aside a ladder that rose into the pilot's cabin. He leaned, but two Vietnamese children rushed onto the plane and grabbed the seats first. Nemo chose another spot on the right side of the bay and strapped himself into a harness. The rest of the squad joined him on that side while a group of Vietnamese filled the other side of the plane, resting their carbines between their legs.

The hatch closed. With a deepening whine, the engines gunned and the plane nosed round for take-off. Nemo felt the sudden surge of powered acceleration as the Caribou first trotted, then loped, next galloped, and finally broke into the sprint that enabled it to pull itself into the air and quickly turn into a steep, spiral climb. He hoped it was only snipers and not rockets the maneuver meant to avoid.

As the plane settled into its flight, Nemo studied the Vietnamese. They seemed almost clones even though their camouflaged uniforms didn't match. He thought about War Zone C below. Somewhere down there existed the mysterious Viet Cong National Command; a Xanadu hidden under triple canopy forest. Only Nui Ba Den, the Black Virgin Mountain, a 3000-foot extinct volcanic cone, rose out of the flat Vietnamese coastal plain, though the coast was eighty miles distant.

The mountain and its legend, that of a Vietnamese girl who threw herself from the mountaintop in anguish for her lost lover, never left him. In his mind her straight, black hair reached the small of her back, setting off her clean, dark skin kissed by only one man. Her breasts were small, erect and untouched. At other times she wore the white, conical, sun hat, silk blouse and trousers of a university student. She often rode a bicycle—always on a rain-washed, paved section of Highway 1, always under the green of cultured banana trees. And when she made eye contact, it was always a glance from a lowered face with only the hint of a smile.

Somewhere beneath the wings and the green, thousands of her men hunted each other in small, hidden groups that spewed fire, blood, and life in a hundred private encounters.

The Caribou veered into a steep dive. Nemo felt his stomach churn as the plane pulled up and bounced onto a runway. With a roar the propellers reversed their pitch and the plane shuddered against the sharp strain of its braking. He heard a sharp, metallic snap and the thin, metal bands that held the steel plankings to the floor broke. Like sticks of gum tapped from a packet of Spearmint, the planking shot toward the front of the plane.

For Nemo, the two Vietnamese children had no chance, and it all slowed down.

A horizontal guillotine, the planks took aim. At some giant metal press sweating in the summer heat of Cleveland, or Pittsburgh, or any one of a thousand other swing shift plants, a man punched out the round holes on other pieces just like these and felt good that his sweat supported the wheels that carried the planes that held the soldiers that save these lands that Jack built.

One sheet of steel caught the boy just under his chin, sliced through his neck, and embedded itself in the firewall beneath the cockpit. His head sat neatly atop the planking. His arms jerked spasmodically, as if motioning for help—not panic, just an urgent motion telling Nemo that if he acted quickly he could reattach the boy's head and nothing would be wrong; no one would notice.

Nemo unsnapped his harness, stood up and tried to steady himself, knew instinctively there was nothing to be done except watch the blood spill across the flight deck. The young girl sat crushed under other planks against the firewall, her eyes dull and vacant as her brother's. And then came a wail from the other side of the bay, a long, vacant sound that transcended even the noise of the plane. A Vietnamese soldier pulled at his hair and the two soldiers holding him. Nemo and his sergeant joined other Vietnamese as they tried to gingerly move the psp off the bodies, but the rumbling, rolling, and turning of the plane mocked the attempts.

Before the plane had stopped turning, the rear hatched dropped and a jungle-hatted major jumped on shouting, "Clear the plane! Move it! Move it! Grab your gear and move! In-coming'll get your ass in a minute. Move it!"

The sergeant changed direction, grabbed a pair of bolt cutters and snapped the metal bands tying down the bucket loader. Someone jumped on, turned the ignition, and the loader burst the bands Helms hadn't gotten to and zoomed onto the runway. The major screamed over the roar of the engines, "Grab this shit! Clear the plane!"

Nemo turned, grabbed a piece of psp in one hand and his rifle in the other then followed the bucket loader. Others did the same.

Outside Nemo dropped his rifle and the psp and ran back onto the plane for another piece. It was wet with blood. He dragged it out of the plane and dropped it onto the runway. He looked at the blood on his hand and felt his face contract. He straightened and held his hand in front of his face. His body recoiled, but he could not look away and suddenly as if swept away in a time warp he was staring beyond his fingers.

What spread before him was not Vietnam, but Verdun. The camp was nothing but a low mound of reddish, earthen walls cased in barbed wire. A battery of 105's fired sporadically from a raw, shallow pit to the right of the camp. Shell and bomb craters dotted the clearings surrounding the runway. To the left trees were chopped and splintered into crazy patterns. In every other direction the forest was the stark gray of winter. Where is the green? He thought. The green of Tay Ninh and the jungle organic and primeval? This place could have been Potter County in January, 19-- it didn't matter. 1916, 1943, 1967, -- it didn't matter.

The major shouted again, "That's it. Fuck the rest." He signaled the pilot. A crowd of Vietnamese was still inside with the two children as the hatch closed and the Caribou roared down the airstrip and lifted itself into the air.

The sergeant gave Nemo a shove. Nemo bent down, picked up his rifle, jiggled it, and sprinted toward the camp.

Prek is the Cambodian word for river.

Xóm Bào Dòn

He watched a dusty platoon of soldiers
ride three five-tons through the abandoned
landscape of a deepening, end-of-rainy-season day.
They rattled past the rusted bulk
of an APC, stripped of all

but its burned-out shell. It lay
half submerged in a paddy. The rosette
of the RPG that didn't get through
splattered next to the molten hole of one
that did. Further on a copse rose

from the dikes—a small, level oasis.
They pulled their trucks in, dismounted,
examined the almost island. Two palms,
a banana tree, a few exotic fruits at one corner,
Bushes in the other. In the center,

a raised rectangle of baked earth,
At the far end of the pad, like adobe block,
A low box of hard, reddish, blackened clay.
Far across the paddies lay the deadly green line
of The Boi Loi Woods. The Hobo Woods

tomorrow they would splinter, plow under.
Tonight, this island would keep dry.
Tonight, the Phuc-U bird would call from the trees.
They dug foxholes into the plot, discoursed
over the pad. One thought it an altar—

pagan's place for worship. Another, how like concrete
they poured for hootches at camp. They admired its hardness.
The moon reflected in the paddies and the Phuc-U screeched.
Then one picked up a flake of charred wood and knew.
He was Nemo. This was foundation.

The altar a hearth for mother, father, daughter.
Chickens had clucked among the bushes.
Gruff, night groans of a tethered water buffalo
echoed from beneath the fronds and a canopy of stars.
The Phuc-U called again from the darkness.

Tunnel Rats

The green day dripped salt.
They lined up--sweaty, smelling rodents,
to swelter inside the whore.
Their bloodshot eyes gleamed fear,
But with a word they clogged
They well-used hole.

At the opposite end of that earthy fallopian
Crouched the other here.
Orders spewed lick ticker tape
And they crawled, then wriggled to meet.

Muffled shots wracked this jungle mattress
And a dream burst through.
Whimpering, urinating, a baby formed.
Nemo dragged it premature from its mother.

JD—FNG

JD Collins, private, Mississippi,
Talks little but his eyes never rest.
Bandoleered, he and I maneuver wary of road.
A crack with a crack, a sniper pot shot
And we hit the ditch together. Luck,
Roll on our backs, a breath. Then a zing,
another report, spasm and metallic ping together:
the pin nicked from a grenade.
JD's grenade tight in his webbing caught.

Five seconds.
Eons roll through his eyes,

> First: Relaxation at the miss, a spitball
> which whizzed past his teacher's back
> who turned, calculated the angle within a chair,
> then long strides that stopped one desk near.

> Next: Logic of the tick time bomb
> secure on his breast, the long
> uncontrolled skid off the edge,
> inevitable calm before the sledge.

> Last: Shock at the face of his end
> fifty years before the rest, no friend,
> no pain's delirium, no mother's call.

> For a child's tiny fraction JD glances for help,
> sees there is none, bulls out a bellow
> and leaps at the sun. The dark grooves
> of his heels mold sharp prints as they crush
> the fine, white dust. He charges,
> no scream of fear—his diploma counts—
> a roar, the lion's vault at the spear.

> Victor Charlie should shoot but doesn't,

gives him his guttural dignity, his belch into death.

JD rushes on, tears at his vest, flings
off his pot, five seconds too short for a strip.

At fifteen yards I should duck but don't.
His diaphragm belts a tune and he ascends toward joy.

In mid-leap—not 4th of July—a clap,
black puff and his whole is pieces.

The Gingerbread House

A/588th Morning Report, Entry: 25 December
"Cause of Death: Total Body Incineration"

The witch fattens our fingers.
Bulls-eye! One hell of a lucky shot,
the armored personnel carrier smolders for hours.
We try to get in—steel burns hot.
Redman, Tojo, Mac—what good were their powers
against a random Beijing flutter,
82 millimeters and a Hanoi stutter,
or that silly millimeter less popped
by Saigon and Big Momma, New York.
A day later (It was our work.) we pry it open—
a mother's oven Hansel prepped.

    In this gloom, baggy-eyed and numb, I dream
    I stand with all the dead of all the wars
    buttressed against my knees. They stretch
    beyond every horizon, a clog that dams
    every stream, no sound but a wisp of fog.

    Into this, a speck, a being appears, baggaged with life.
    He finds slowly between each body a slight seam,
    enough for a gentle step; he passes
    none without his touch like a mother's
    left-at-home kiss upon her child's sleep.

    As this tattered witness approaches,
    He surveys the wreckage of twenty centuries
    and turns, faces me without envy. His eyes
    brim with knowledge that never can spill.
    Rounded by history's brutal passage,
    his shoulders bear every splinter festering in each.

    With a gesture more ancient than my race
    he cradles my hands. Flesh on flesh passes,

a kinship, a recognition one could this tumor,
this folly, this arrogance transform
into the holiest shrine. He moves again
to the corpses. They soften.
I begin to see seams and touch.

III.
R&R

Minerva's Music Room
Holywell Street

During the concert in the cold-stone hall,

stolen from her moor, locked behind notes,

stripped of her lance, Diana straddled the lunar cycle.

If she had leaned closer, inspected the bars

the intense moment would have come;

the ancient rites returned. Again

she could hunt by full moonlight

the fox. Spear balanced left,

her cold shadow trotting between

the spare shrub and wild grass

transfixes the red-eyed fox.

One swift motion chops short

its startled bark, fells it.

An owl screes once from an empty branch.

She boarded the plane.

Hong Kong

R&R, I&I, Rest and Intoxication, Intercourse and Recuperation, it didn't matter to Nemo so long as the food was good.

He looked down from the plane as it circled Hong Kong Harbor. The civilian airliner which a week before had crashed short of the runway, lay still in the shallows of the bay like a silver albatross with a broken back.

The city did not seem to notice.

Kowloon

As Nemo left the plane, the cliffs of the Kowloon Peninsula rose behind the city proper greater than any wall history had taught him. So this, he thought, is the judgment seat where East and West finally meet. It did not matter that the political border was miles beyond the cliffs; he could feel the British Empire was itself out against the dark rocks.

For two days Nemo soaked and bathed in the steam and heat of bathwater in the hotel until his crotch rot dried and disappeared. He ate Kowloon beef for breakfast and beef for dinner, bought ivory from the hawkers in the markets, pretended to be British to American soldiers, American to British soldiers. Rested and recuperated, he began to search out the intoxication of intercourse.

At nineteen it felt good to strut. Diana was flying in.

Diana's Glass of Water

Light is the lion come down to drink
—Wallace Stevens

Between them rose an almost empty crystal

Holding last drops sensual as wine,

Enough to create a heart's oasis—

A green in the ancient tawn

Of the driest Kalahari—

A sniff that draws the bushman

Trotting forever after

The scent of soft bathing.

Victoria Island

Diane bred beauty despite contempt.
Her playgrounds smelled of fresh breezes,
Dark hair, green perfumes,
Places children know and lose.

They met in the legendary garden
Beneath the soft crescent of Venus
Where apples and snakes ripen irresistibly.
Nemo breathed her fragrance and ran.

Macao

At the Blue Wall Lounge Nemo leaned back, placed his cards face down on the table. He squeezed an oriental in one arm while his other fondled a drink. Smoke rolled on the heads of the crowd.

Diana snaked her way through the mumbling drinkers and faced him across the table, "What's in the cards."

He met her eyes, "Flush."

The table stupored and grimaced.

She stared at the girl, then back at him, "You fold."

He leaned forward, kissed the girl on one cheek, whispered, "I gotta split," and stood.

He elbowed a path through the slurred crowd and just before the two vanished into the haze, Nemo turned to Diana, "It was a long hand."

"I know."

IV.
Thein Ngon

The Wound

Jesus Christ must think we're nuts thought Nemo as he watched from the back of second squad's five-ton dump while the drab convoy ground its way into a small, flat clearing pushed by Rome plows from the shell-torn wilderness. Oberton, a sandy-haired, stocky, soft-spoken North Carolinian, watched with him. The late sun streaked through boiling dust as one truck after another packed into a tight cluster of milling soldiers and equipment. Sporadic small-arms fire would crackle then die in the bush a hundred meters beyond the clearing. Nemo jumped off. His shirtless back glistened with a thin covering of sweat. Oberton followed and they ambled toward the rest of their squad standing by the side of the road.

Nemo's drooping gait knew well tropical war. He'd been in-country eleven months, forty days from rotation to the States. Spending that last month in the field gnawed at his guts. He wanted to go home. He wanted to go anywhere it was safe.

Obee spoke first, "Did you see any infantry?"

"No, and it makes me feel like bait." Nemo spat and pushed the thought from his mind as their sargent walked toward them. They waited instructions.

The sergeant spoke with a nasal twang and pointed along the dirt road, "We're to dig in here. Bucket loader'll be over to clear some bunkers. Put two people in each. One on, one off. Don't ask no questions 'cause I don't know no answers."

Nemo listened to the grumbling and pictured nights with sleep cut in half, "How the hell are we supposed to build a damn special forces camp if we've got to stay up all night?"

Helms grinned and ambled off. He turned and looked over his shoulder, "Wait 'til morning. When they find all you guys snoring, they'll make a change—unless Charlie gets you first."

Nemo surveyed the line Helms had pointed out. He knew, as veteran of the squad, he had first choice to locate his bunker. The land was flat. The road continued straight unitl it merged with the green jungle. Three miles (five klicks?) up that road lay Cambodia, distant as Tibet, forbidding as the moon. A second dirt road formed a junction to his left and cut straight to the horizon. He knew, two miles (three clicks) down that road framed in unscarred bush, also lay Cambodia. In front of him, between the two roads was an acre of half-cleared forest and groups of Vietnamese with their advisors. They dug their own foxholes and bunkers. Jesus, he thought, I used to walk that far to school. Now I wouldn't make it a football field beyond this clearing.

Nemo pointed to the junction, "Let's dig here. Obee you're with me."

They grabbed their gear from the five-ton and walked toward the spot he had pointed out. Nemo opened his lungs and the green of the forest seeped through the dust and filled them. Daylight outside the base camp was freedom; new lands, new lights. The trees here hadn't been recently defoliated. Their leaves grew curved and ripe, full with natural wisdom. The war had been absent here for some time. He felt a pang of guilt and sadness—the same sadness he'd felt as bulldozers cut the new highway through his father's Pennsylvania fields.

Obee's voice drew him back, "Did you notice how smooth the road was after we crossed the river? No craters either."

"Yeah, somebody's keeping it up. If it were us, they'be wider. Shit!"

Obee responded, "I had bad feelings about this thing when they called it "Operation Crazy Horse" and started it on December 7. You know what I feel like now?"

"Custer?"

"No, Custer's men. Custer is back at Tay Ninh."

Nemo groaned, "Christ, I wish we had infantry with us." Infantry had been a mother to them.

Anytime they moved into the field, infantry ran interference. This time the company had been sent out with Vietnamese irregulars advised by Green Berets. His gut tightened as the shadows blended quickly into twilight.

The headlamps of the front end loader grumbled toward him, and he forced himself to forget. "Here," he pointed to the spot he wanted dug out. The loader paused, revved its engine, jerked forward and rammed at the red earth. It clawed out the dirt and again he was home watching a dozer dig another cellar hole in the developments sprouting like corn all through the fields. It peeled back the rich earth that once grew corn and fed Guernsey. Next would come foundations, then ranches, macadam, sidewalks, wires, and malls. The new children would know no difference.

He brought himself back to the loader and looked up the road toward Cambodia. Two flashes and then a pair of pink lines streaked toward him. The first flashed into a trailer and burst into pink fire. The second slammed into the bucket loader. It bucked, settled with a sigh, its driver vanished. Nemo stood transfixed at the soft colors until a third rocket exploded into the soft dirt and he felt the heat singe his hand. He flattened himself. Grass smothered his nose, dirt kissed his lips. God, he thought, not now. Please not me.

A mortar burst beside the shallow ditch he lay in. Scraps of hot metal spurted overhead. Another mortar burst farther way and then the clearing exploded into sound. Nemo raised his head. Glowing tracers rocketed in every direction. Someone was firing a machine gun almost straight into the air, its tracers bouncing crazily off branches. The smell of gunpowder drifted through the strobe of bursting shells. Parachute flares gave the perimeter a stage-like appearance where only red streams of tracers spoke like incoherent actors. Then, with sudden clarity, Nemo realized if Charlie decided to come through, they had no chance.

He looked over at the loader, thought, lightning won't strike twice and salamandered through the soft dirt toward it. When he reached the loader, he curled inside the gaping metal mouth and scooped dirt in front of him until only a firing slit remained. A flat-bed truck lay directly in front. He fixed his bayonet he felt a pang of guilt and sadness.

And slid his rifle onto the slit. He knew he would shoot the first oriental head that showed above the trailer of that flat-bed.

Minutes? Nemo felt like hours but understood it was probably seconds when a black silence returned to the clearing. Another flare popped and cast an eerie slit of light into his hastily made foxhole. Against the back of the loader's metal bucket he noticed a dark liquid streaked down the metal. Its reddish hue turned black as the light of the dying, flickering flare danced against it. He wondered about the driver's final thought. Did he see it coming? Was there time to curse before oblivion? As he looked at the liquid he felt a bond between himself and it—a bond that hadn't existed moments before. Nemo slowly moved his hand toward the streak. He touched the stain and suddenly felt absurd. It was oil not blood.

Nemo pushed his hand into the soft dirt and then pushed the dirt away from front of the bucket and slithered out. He crouched, "Is anyone here?"

"Shut your damn mouth!" came a hoarse reply from underneath a nearby five-ton.

It was Obee. Nemo's dusty mouth split into a grin and a sprinted toward the voice. Obee had parked himself behind the rear dual tires. Nemo's voice strained, "I think it's over," and he gave Obee a hand to pull him out. They both slumped against the tires. "Man, I thought we were done," Nemo said.

"You're not the only one," Obee picked himself up and brushed off the dust. Somewhere in the night men barked, machines shuddered, and the compound came alive again.

Helms walked over, "Anyone hurt?"

Obee inspected his trousers, "Nope."

Nemo interrupted, "The driver. I think he got blown away. The RPG hit dead on the loader."

Helms stiffened, "Nuts!" And walked away.

The Beagle

Without the loader, they would have to dig. Obee jammed his entrenching tool into the ground. Nemo took out his and began to help. His muffled grunts bounced inside the quickly deepening trench. He pictured the bunker gradually taking shape. It would be about four feet deep, seven feet wide and ten feet long. When it was finished, they would cover it with PSP and sandbags. Rifle ports would face the junction and an entrance would be dug off the side.

The two lost tack of time as they shoveled. Nemo's earthen wall was just about square when he realized Obee had stopped digging. Nemo straightened up and looked directly into a pair of crisp uniformed legs. A beagle sat behind them. His gaze slowly followed the tailored lines upward. They were not jungle fatigues. He knew when he saw the small, smooth hand holding a carbine that by the time he reached the figure's face, he would be staring at an oriental.

The CO's warning flashed, "All friendly irregulars will wear yellow-and-blue armbands on their right arms." Nemo examined every inch of the olive drab shirt, the bandless left sleeve, the bandless right sleeve. He fingered in the dirt behind him for his rifle, but he knew there was not a chance in this hell that either Obee or he could get to their rifles in time.

Nemo looked at the man's face. The Vietnamese grinned. His black teeth gleamed. For a long moment the three stood wordless and smiling, mutes in an asylum.

Obee broke the silence, "You all want something?"

The Vietnamese talked in a pitched babble.

Nemo looked at Obee, "Jesus, why can't they speak English."

"I'd be happy with French," Obee stated matter-of-factly.

The man spoke again. His second wave of words washed over them as meaningless as the first. Nemo thought of the old peasant women stringing grasshoppers to a blade of grass, spittooning blood-red beetlenut juice onto the ground from between their black teeth. Grasshoppers might be a delicacy but his stomach squirmed at their coal black teeth and balked at the live what-looked-like rats; jaws pierced by leather thongs tied to stakes. The art made refrigeration unnecessary.

The brown-and-white beagle sniffed the air behind the Vietnamese. It resembled the squad pet at Tay Ninh. He wanted to strangle the gook who had stolen and eaten his dog. He thought if only I had kept closer watch over him. A lump clogged his throat.

Obee turned, "You think he's friendly?"

I don't know." Nemo pointed to his arm, "You have arm-band?"

The short man excitedly pulled a yellow-and-blue cloth from his pocket. Nemo stopped drumming his rifle and unpacked a C-ration from his pack. "Here dog, you want some food?" The beagle wagged from its neck back. "Obee, give me your opener." Obee tossed his P-38 to him and Nemo opened the green container of stew and gave it to the dog which greedily gulped it down.

The Vietnamese repeatedly motioned from himself to their bunker. Obee finally understood, "I think he wants to share our house."

Nemo had forgotten all but the dog. He rubbed the pup with both hands, "Give him a shovel. If he wants to dig he can stay for the night." A soft, fluttering sound fractured his comment. "In coming!" he yelled and dove at the incomplete bunker. He tried to submerge in the deepest part of the trench as one heavy kwump after another resonated through his head. I hate them he thought. Why me? Please miss.

Two howitzers boomed and the sounds of bursting mortars merged with small arms as the camp exploded again with noise. His fingers clawed into the earth; he wanted to join with it, feel its richness squeeze against his palms. If only he were earth he thought.

Nemo felt something nudge under his legs. The thought of the Vietnamese trying to use his body as a sheild turned fear to rage. He screamed and grabbed at the intruder, but it wasn't the Vietnamese; the whimpering beagle pawed at the earth under his legs as if a last great bone lay buried there. The rage vanished; he cradled the puppy in his arms, rolled onto his stomach and they trembled together.

Soon eight-inch shells bumbled in like boxcars and erupted in the forest. Nemo looked up to see a Phantom jet scream in at treetop level and drop napalm. The boiling explosion turned night into day, life into ash. The explosive symphony crashed out its finale and silence rushed over him like a wave washes the sand. He slumped into the ditch. The beagle had stopped whimpering and Nemo filled the dog's forlorn brown eyes with all the heartache he could no longer contain. Cinders and craters he thought. We are creating a moon.

He rolled into his back and stared at the flare-studded night. The beagle snuggled against his armpit. The silver Phantom, free of its napalm, slowly circled the compound. Near stall speed it seemed to float like a spaceship. Flares lit its inward side. Its wing lights blinked slowly. The plane kept circling like a remorseful father who has whipped his child into obedience. Fatigue worked on him like a drug. The Phantom's outer half was lost forever to the night. He mumbled to no one, "No plane can fly that slow." He remembered his plastic models, tiny tanks and soldiers, the mock battles with his brother long ago. He once told his brother that he loved him, now he was glad. No plane can fly that slow into mist; he dropped asleep.

War Dog Memorial
Barrington, New Hampshire

Next to the Veterans' Stone sits
an acrylic German Shepherd, panting
for command. In front, a small flag,
a whittled stick, and two large feathers
tied with leather which lift in the breeze.
Like an illusion, the evening summer
haze simmers about a drooping sun.
A distant lawn mower whines like a gnat.

Draped about the Shepherd's neck, dogtags:
Colonel, Roxy, Satan, Cochise. How easy
they trained—anxious, eager—man and dog,
while others strutted the suited wire of duty
and looked down, knowing the dogs
would sense too late, in that sudden crash
of automatic weapons, recognize
like their men all had been forsaken.

The wind turned on my cheek. I saw
Old Barley lift himself at each
late-night mission, wait by the door
acknowledgment and after plod to his bed,
circle three times, and drop, one great sigh
escaping as he drifted into sleep.

Flashbacks, Kunitz's 90th Birthday Party and Reading,
1995, Worcester, Massachusetts

The wagon's still wagon, still red,
circling this hundred near, this coming of comet,
this metamorphosis into word.
Was it the mother that slapped him hard?
Sting so hot it vaulted this soldier back thirty years . . .

     where he too quickly recovers
     from the slap of shrapnel.
     A coy girl allows him to toy.
     Will there be days enough? At nineteen
     the body's quick, hard, apt as rifle.

     The morning in-spite-of-himself-he-heals comes.
     In two hours he'll chopper back to war.
     She unbuttons her blouse, breasts
     small fruits, nipples dark, tangy as pears, . . .

golden in a Woodford Street garden.
Kunitz, pelagic mortal, closed his book,
tears in his eyes, *"Who are we? Why are we here?"*
His cake is crowned with an open volume,
its leafs, icing thin as silk—like her silk . . . .

     The girl gathers herself, stands tall
     then lowers her black trousers.
     Time arrests itself. Flaming scars
     lace the skin of her crotch,
     her inner thighs equally slashed—

        Bayonet or suture,
        Shrapnel or razor:

Which side? Enough to know the fathers.
     At last he meets her gaze,

(I'd like to think he held her, cradled her through)
        but he turns and leaves. . . .

Who are we? Why are we here?

From a tree at My Lai

"Moments Before the Massacre"—R. Haeberle photographer

The old woman knows. She fingers her kerchief
like a babushka kneads her rosary,
like a nervous Baptist plies his Bible
whose soft, black cover links the days of their past
to this night-in-day present. No priest will stop
the gunship god as he hovers certain, angry,
seething spit, and splutters, "Mother Fucker
I need flesh to finance my church."

In the background, one daughter fingers her silk,
this her final marriage. Her husband
took his vows earlier, rehearsed his blank
scream at the adulterous battletank
before its beehive stung him silent, standard.
Purse not that he was better; he too had opened
Many beds to other men. Just, he killed, not photographed.

She hips her son, balanced, waiting the bus
that will jolt them out of time. The boy,
curious, stares at the finger happy grunts
about to trigger the toy which builds his tomb.
His school will end before he learns.
He will not feel his body's earthward drift
but think, as slugs twist and squeeze and lift
his frame, "How dumb. How dumb. How dumb."

Voyager, I had hopes you were different..

The Fourth Day

Despite God having quit this place days ago, a humid morning arrived on schedule. The endless Chinooks dropped in and out with their cargoes and dust. The squad sargent walked over to Nemo's bunker. "Ok, we pull security while headquarters tries to pull a tank out of the mud Any questions?"

An armored unit had arrived yesterday and already had lost a tank and APC. Nemo had no questions, picked up his rifle, flak vest, and an extra magazine and joined the rest of the squad as it walked single file out of the perimeter and down one side of the road. They hadn't gone far when they came upon a tank hopelessly mired in the gray swamp of a muck-filled bomb crater. The squad dispersed along the side of the road.

Nemo walked a few feet into the undergrowth and left the tank rescue to headquarters. He picked a grassy patch near a fallen tree and lay down. The grind of engine and gear soon blended into background.

The impenetrable forest steamed within yards of his lair. Nemo carefully arranged the grass, making himself as inconspicuous as possible. He peered into the green looking for movement.

A lizard on a tree near him flicked its tongue. Its whip-like tail trailed behind orange and blue scales, its cold-blooded muscles poised to snap. Following the branch, Nemo discovered the mini-monster's intent: a giant grasshopper lurched unconcerned toward digestion. The lizard grinned, blinked an unbrowed lid and pounced and quickly crunched grasshopper into swallowable chunks.Satisfied, the lizard casually looked over his shoulder at Nemo and waltzed into the undergrowth.

Warm and alone, the day swooned in the tropic sun, but Nemo could not forget the lizard. Again and again he replayed the scene. Vanish! The thought hit him like a sledge. The lizard, the grasshopper, he must get home or vanish as they. He looked back at the tank. It had turned the water thick as vomit.

The tank never made it out of the crater.

That afternoon, the squad returned to the camp. His bunker mates walked over to headquarters bunker to fill sandbags. Nemo ambled alone to the bunker. The late afternoon doldrums quickly mellowed the raw landscape. Hidden birds cried their lonely stories against the red sunset. Gradually they too lapsed into silence. He began clearing trash from the low, sand-bag topped, bunker, carrying old cans and ration boxes out the side entrance and onto a pile outside. He hummed a faraway tune with no particular melody.

He turned to go inside, when something whacked against the back of his head. Anger flashed at the thought of some idiot throwing stones because he hadn't filled sandbags. He grunted, groaned, and pitched into the bunker. Why am I grunting he thought, his reflexes far ahead of his thoughts. It wasn't until his hands returned from the back of his head, red with blood, that he realized what had happened. He had heard nothing.

Time slowed to a stop. He saw Obee and Corky rushing toward him in slow motion. They reached him, each grabbed an arm and steered him toward the aid station. Howitzers returned fire and for the first time since the attack, he heard noise. Another howitzer cracked and all self-control vanished. He broke from his friends and spurted like a rabbit into the first aid bunker. Two medics grabbed him and flattened him face down on a table.

The medics worked passively. The room glowed dimly and one remarked, "Send him in. I think he's punctured a lung." His platoon leader ducked in and held his hand.

Nemo wanted to say something, but no sounds other than groans escaped. His mind seemed separated from his body. He thought, am I dead? Two soldiers placed him on a stretcher and covered him with an olive-drab poncho to keep off the dirt. He felt himself lifted into the dark.

Like lightning, muzzle blasts of howitzers reflected inside the poncho. He could picture the pellets of high explosive whiz out to crack in orange light against some innocent tree. The stretcher rocked with every step of the bearers. He wondered if they were scared and gradually a fear of falling replaced a fear of death. From beneath the poncho he watched a world framed by rubber.

A medivac swooped from the dusk. Stretchers and bearers popped from the ground like gremlins invading a Satanic dance. They stacked their sacrifices on rungs across the chopper belly. The stretcher before him was legless. Then he felt himself lurched upward and fastened to the top rung. He looked out the open side door of the Huey. The medivac jerked and he was airborne.

Seconds later he looked again out the door into two thousand feet of empty space. The moon illuminated the jungle. His fear of falling lost itself in the beauty below. Silver mists drifted through the trees. To the east Saigon glowed like a jeweled crown; its river snaked toward and under him, black as blood in the night. Near the horizon a firefight sparkled in silence. Red and green tracers searched each other like ray guns from some out-world festival. A battery of shells burst one by one in a neat, silent line. He wondered if that was his compound or some other lost fight in this torn country. Still the great immensity of the land, rimmed by the moon, lay silent and silver.

What is this place he thought. Now I am vanished from it. It still breathes yet we cut out its heart. No more. This land—I am no doctor to slice it up and sew it together. The dim lights of a base camp appeared out of nowhere.

The medivac circled to a landing and medics swarmed at it like pirates looting a frigate. Hands whisked his stretcher out and onto the ground, then released it. A voice shouted into the roar of the chopper, "This one's dead."

V.
45th Evac.

And God Saw That It Was Good

Once people, no more they die except to us. They won war but could not win peace, a grace war-torn peasant found. No breath copulates like dying, no mate faithful as death.

Dead! Dead? The words ricocheted in Nemo's brain and then he understood: The poncho—the rubber poncho! They always wrapped the dead in body bags or ponchos. I am not dead he thought.

He tried to yell but still could make no sound. He tried to move his arm but it would not move. His body thought for him, and he kicked one medic in the shin.

"Like hell he's dead! He just kicked me." The words emanated in Nemo's ears as if from heaven. Instantly he was lifted, rocked and run past the sign that said 45th Surgical and inside. They dropped him onto a table. One medic stripped off the poncho, another held him upright, a third slit his bootlaces and belt, then ripped off his trousers and boots. Nemo thought only of the waste and wondered if he'd get his boots back. He tried to speak but his voice remained locked.

A middle-aged woman walked over, her face wrinkled in contemplation. "Turn him over," she commanded. They did. She prodded, "Well sonny, they certainly peppered you, didn't they." Nemo thought and imagined he nodded. She pushed and squeezed, "How the hell they missed anything vital is your lucky charm. You seem in pretty good shape." She turned to the medics, "Have him x-rayed."

They wheeled him to an x-ray machine and when they had finished, they moved his table off to one side of the operating room and moved on. Nemo sat up slowly and began to unwind.

A chaplain, clutching rosary beads, walked over, "Do you have a religion, son?"

Speech finally returned, "You're not giving me the Last Rites are you?"

The chaplain chuckled, "No, no, they'll have you fixed up in no time. Where are you from?"

"Pennsylvania."

"No, I mean what unit?"

"Sorry, A/588th Engineers."

"You want a smoke?"

No thanks, I don't smoke."

"How old are you?"

"Nineteen."

There was a pause.

"Well, if you need anything, I'll be right over there," the chaplain nodded goodbye and moved on.

Nemo propped himself on his elbows and surveyed the room. All operating tables were filled with Vietnamese, each attended by nurses and surgeons. He examined the bare feet of an unconscious soldier lying on a table across the aisle and wondered how he had joined their army with a club foot. Then he realized the heel had been neatly sliced off by shrapnel. A surgeon frantically pushed and kneaded the torso muttering, "I can't stop the bleeding." Each time he lifted his hand, a flap opened on the soldier's breast and blood poured like juice down the chest. A nurse held compresses over lesser wounds. The partial heel trickled unnoticed fluids onto the table.

At the next table, surgeons suddenly stopped working, disconnected the plasma bottles and motioned to an orderly who wheeled the body away. The table squeaked as it rolled out and another rolled in.

Another medic brought a sheet over to him. "Do you thank you can walk?"

Nemo nodded.

The medic continued, "They checked your x-rays and found no apparent skull or nerve damage. Everything hit bone. So, we're sending you across the street to get sewed up." He wrapped Nemo in the sheet and pointed toward a building across the road.

As Nemo walked out, he looked back at the fragile room and those in it. They looked young, like boys. His now club-footed friend across the aisle still bled and the room remained quiet like a classroom at test time. He pushed through the double doors.

Outside, the night dreamed. He wondered where the noise had gone, the confusion, the breath of straining men ? Here the quiet was of a well-kept suburban street. He blessed it—no guns, no trucks, no fear in the dark. The silence bubbled from within and welled down his cheeks. I am no doctor, he thought. This infected land, my infected people have no doctor. But yet I exist. I will go home. What else can I do?
Nemo disappeared into the dark.

Error

In the dawn,
after Matty-16-Mattel
and A-K-Plenty-47
had played their tag and groans
disappeared with the dark,
light filtered the moans.

Survivors, dusty beasts
bent on extinction
lifted themselves from the earth
and stumbled away.
They were not the first.
One old man wandered

forty-five years before he found a wall
to help steady both barrels,
while his toes toyed
with the triggers
then exploded the roof of his mouth.
A second dreamed the brains

of a thousand gunners scooped
like sauce onto the plates
of a thousand runways. This planet
insisted he splatter his guts
across the hood of a Ford.

Shuttle Diplomacy

I. Preface
In the outness of Space
that belly of darkness, that absorbance of light
where years empty like seconds,
lost sailors cry
and swollen puppets lie.
Where nose nor echo ever responds,
even dust becomes a sought neighbor.

II: Liftoff
Ignition slowly lift
this perversion of hell
untethered from the land.
Rise cutting edge of stars,
flame through the dusty band
of breath bound to the settled mind
and take me home.

III: From the Vacuum
…Beyond, against the black velvet resolves,
a blue-white marble dewdrop floats
and despite conflicted tides revolves.
Balance and grave like God and Job
against this dark slate orbits your fragile globe.
Bipeds bathe your children. Show them your knees.
Their star-wide awe accepts gliders birthed
by ocean and blast, cradled in sand and breeze.
Dreams fill their voids. They wake not to cry
against an end but to urge the next to try.
Steer through the terrestrial mantle;
let daughters be Noahs:
let sons be arks.
Only then count your self-worth
and sense yourself free
to know Magellenic visions
deeper, far deeper,
than Columbus's sea.

VI.
Going Home

Nemo's Return

The road, a black serpent striped white,
scales the desert coast inward, and he,
boatsick centurian, lurches ashore to drive
this reptilian hump toward its den.
Telephone poles like intrusive pikes
flash past his window—a line
of crucifixions that carry
the coarse voice inland.

Home is full of strangers; daughter grown weary,
hot for the city; wife shackled by childhood,
a barbed wire ring of domesticity;
lover retreated into the cosmic hole of security.

If only this snake coiled into a green, an oasis
where all the world's flowers gather in hush,
he'd solo differently—not so rampant
like the nib whose ink has dried
leaving only coarse scratching on bark.

The snake strikes left. He stands
on the brake, skids this sliding rush
to a halt at the edge of a ledge
and is slammed back into the seat.
In the peace that follows, he calibrates his choices.
All leave him sole proprietor of an empty estate,
learning's executor stranded on the vacant
dust of the shoulder—a lane leading nowhere.

The radiator boils over. The battery shorts.
The gasoline line locks.
She . . . She . . . She . . .
Where does one lie for a woman?
Oh he could kick the tire, shove his fist
deep into the fender—all
has been tried. Instead, he squints
at this dry landscape, squats
until his boney haunches ache
and fingerpaints the sand.

Night Walk

In the northern nights before Christ's
cigars and gifts, the yellow half moon
floats above winter crusted branches,
an oriental lantern that hangs hints
of encounters to come, a throat of themes
that pours pattern of shadow
over acres whose breathe steams
silent toward the dipper.

I walk the solitude,
and hear Diana's voice rise
in a sonata of promise.
The lyrics her now, the melody her youth;
the ache between what is and what could ....
The song carried clear in the air.
Nuances nested into sprouts
tight and brittle in the ice
that crusts these hills.

She sings not for me or man
but for endings she can not touch.
Her lips shape sound for solo hearts.
Oh woman—softest instrument—loneliest tune,
                  I am ear for your voice.

The Cutter

He imagined himself God's image as he pushed
his rotary mower. He smiled when he turned
and surveyed the lawn he had carefully laid.
As a boy, he cut the town cemetery and learned
its stone history: The choicest blades, trim of death,
and a breeze that eroded the neck of a cross.
Then, after rain, steam rose off granite
above the catacombs where bone and memory jumbled.

Now the wilds mock to the lip
of his lawn. Tall, intolerable sprouts
will not stay level like stone.
Here the mourning dove calls from beyond
the reach of the forest where trees lock
out the sun. Here the scent of pine turns
the nape of his neck stiff with scruff.
He snarls at what soon his teeth must cut.

Figure Skater

Diana skated into spring
on ice she never trusted, but it held.
Now while sun poured over the snows
and the melt sang its way to the drains,
she reached for Nemo in his car, squeezed his hand
then turned back to the slushy bank
and leapt like a Thompson's gazelle,
in nature's un-choreographed ballet.
After four symmetrical waltz steps
she leapt again, toe loop of hope,
over the far-side bank.
But he had vanished through the post office door.

They loved in both times
though neither could quite remember,
but with that forward leap,
trialing toes shaking off that misty past,
she sliced what ribbons remained.
The only axel which kept him contained.

In Pursuit of the Loon

The marsh lay grey and still.
Mist drifted like thin smoke.
Swamp grass grew grey and silent;
A few stunted tamaracks poked their black beard
Through the soundless contrast of level color.

"So Lonely ..."

The loon's slow echo reached me,
Hands that cupped my ears in sobs,

"So Lonely ...."

My grey raft swelled silent in the slate marsh,
Each oar carefully placed and drawn
So that I moved like leaf in stagnant pond.

The loon knew but stayed
Always at the edge of vision.
Like a ghost he beckoned 'til I closed
Then submerged leaving an image vanishing like vapor.

Seconds primed ages, and he emerged without a ripple
In a more distant place. He soothed, a siren of loss.
His black-and-white-lined throat seemed
Stark against the sifting curtains of grey.
His eyes glowed red as he called,

"So Lonely ...."

It echoed to hills hidden in mist and returned;
It filtered the shore in riffles that wash and restore,

"So Lonely ...

So Lonely ...

So Lonely ...."

Granville Reservoir

The riffles pass,
wind, liquid, stone—they lick and go.
The breeze passes,
a child's heart whisper: Past wood, past snow.
Mine aches to pass like wind and wave,
sift the tall grass and know.

        Where does it come?
        Where does it go?

The waters know. Silent and steep
the green-lawned earth knows.
Daisies, uncurling their yellow seed,
spin the song through their petticoats,
but I am deaf.

        Liquid she knew and floated
        over the spillway.

My Sea

I

I find the seam

between seedy and sublime.

Like the crack in the aquarium

appeals to octopi, I slither through

parroting the background: red on red

white on white

blue on blue

disparate gulf between

what I was and what I feel.

Tentacled mollusks know that gulf.

They stream in its shadowed depth.

Their eyes become me; I their apprentice,

and my brine, their league.

Within my veins lurks the squid.

Jonah and Pinocchio know

the belly of the whale safe

from all but the reach of the past.

And as I, cold as beak, turn;

my digits lengthen and stretch

into the squid that strangles the sperm.

II

For the fortieth time I sit and wonder
like Noah's rain, ignorance abounds,
but Noah had an ark with two of all friends;
I've a brain for a barge and six leaky senses.
I've signed a crew ripe for spoil, talk, naught.

I sail a sea foul with scurvy, salt, rot.
This tramp turned ark, turned *Bounty*
and I turn Noah turn Bligh.

    "Mister Christian!"

    "Aye, captain, all Christian
    in name. but breathe
    independent and the names,
    yours and mine, congeal
    in dank, infested holds."

    "Fletcher, I am not judge.
    My crews all sicken the same.
    I am executioner."

Noah had his faith,
Bligh had his ship;
I have my doubts.

III

And with a yelp torn from time

I cast myself toward shore to find another way.
I sail past the wrecks at Wiscassett:
sun-bleached ribs warmed by the August morning.
No angler, man or beast, loots the tide-worn sides;
no bather, skinned or scaled, questions *Little* or *Hesper.*
They are ghosts and witness. I leave them knowing
They would float to haunt again.

IV

He lofts his keel to shelter and new timber.
Sure enough, in the end of autumn
gloom while beached from the hull of war,
Eve's loom, bleeding doctors galore,
Adam's rub, he waits another tide to salvage his bones.

When out of the Wiscasset mist and mind floats
the four-masted hulks. They beckon him to the flats,
"Why not rest in this mud? When you venture again,
the hurricane will strike, shred canvasses,
split these masts, flood the hold with salt
and force your old craft to loosen stays
and like *The Dutchman* slip beneath the breakers
you call togetherness."

V

Nemo packs his pipe with spiced tobacco,
"Perhaps," and lights, "but not this day."
And with a puff born from all life
blows the spirits away and waits
the sound of smoke, a distant fife,
and creaking wood, a tide to sail the bay.

After *The Nutcracker*, Nemo's Letter to Diana

Early darkness cops the holiday night
and the lateness of this bus, a cocoon
of soft green light flourishing north from the city.
Inside, a man's daughter curls beside him in sleep.
His arm snuggles her Snow queen, her Sugarplum... .

His other hand cradles a pen that scratches
warm distractions blended in the whine of diesel,
"I watch Christmas lights flash by: Their hopes, their cost, their loss.

The words string circuits and colors leaping
rooftop to rooftop, every chimney full with gift.
"My thought turns to you—the only other
who could nestle in my night cloud heart,
round, wide as the white moon tonight.

But—

"You rest distant in another's night ....
Were I Faust I'd become that moon—
sparkle about your sleeping shoulders
in a silent comfort of lace.

But—

"I'm not lunar. I am man—if that—
holding a settled child while this bus
hurtles us deeper into the night. Still.
This gift spreads ink, fleeces the chin
off that fellow in the moon.
I pray you sleep soft."

VII.
Burying the Oar

Lonely Toes

"Lonely toes, I have lonely toes,"
His daughter ployed as she stretched

> on the backseat then propped her feet
> between the Chevy's front seats.

> In that tired Sunday evening drive
> he took one toe market, one toe home,
> and she relaxed with a moon smile
> that filled him with wonder and rage
> because his love's a bone stump hacked by a dull axe.

She wiggled and he squeezed past

> the staccato chop of bullet and blade
> that stacked boys into medivacs,
> their splintered knees gristly white
> protusions through the strands
> of ligament and tendon that once hooked them
> to calves and shins still laced into boots
> left crazy quilt on the ground.

She squiggled and he clawed further back

> to the burning oil of a mother,
> doped with morphine, missing her breasts,
> half her weight, her tallowed head dropping
> clumps of black hair in the best radioactive style.
> Still she could smile recognition to her child
> while filthy surgical hands pushed him,
> pushed him out of her room.

> He screamed into the present,
> clung to a daughter's toes
> as if they could stomp,
> as if they could kick
> and boot him, boot him, boot him
> out of his pasts.

Upon Discovery of a Mammoth Hunter's Grave in France

In ticks before a clock bisected man
From ice, before his senses lost their tone,
A child skipped till death required she own
A different dream beyond her parents' plan.

She listened, children do, and wrestled death
Alone. Her mother dressed her soft in skins.
Her father's gift: garland, horns, whittled pins.
With strings of wind, they clothed her last with earth.

A hundred thousand winters melt; I feel
her flint, her hair, her kin. Computer chips
divorce my daughter, death from daily health.
A punch cross time, cold hands, a wind-knocked reel
that catches breath, my inner horns, my slips—
two children's gifts: their lives, *oh God*, my wealth.

Cleansing her bedroom

Like the Milky Way's ethereal glow
behind the meridian of the inkblot moon,
daughter's things spread in a band
across the center of her room,
and in the corners other galaxies,
piles of dust and distant things.
It falls upon me to play god.

I flip back forty years to my own
clapboards, room, evolution
from photos of ballplayers and generals
to sheathes of notes and books.
My father too had his line to god
but never commanded because my room
lay unheated two floors above
places neighbors, thin aliens, might show
and because no mother softened the blow.

I watch her collapse collectibles
into a pile dense as a neutron star
which under its weight, might collapse to a black hole
or as easily burst into light that burns both me and the trilobite.

Other God that plays beyond the shadow,
litter of Bibles, and paycheck gospel,
speed her from the sentiment and sediment
that makes me fossil, red-shifted and slipping below.

Horses

Late August, if machine does not intrude
there can be noon, a grazed field,
weathered fences surrounded by axle straight pine,
crickets willing their occasional chirp.
To be child at that moment aches beneath my gut.

To hear child across the thistle, see her
ride bareback, her pair of mares
an obedient tandem of joy
bubble like an artesian spring
soothes joints burned dull
with arthritis not of the bone.

Memory is too much saddled. Better to live
bareback, enveloped in the horses' scent.

Two Love Poems: Christmas and The Other Season

1.

Christmas

—over Haydn's 6th.Symphony, 2nd Movement, *La Matin*
and Thomas Coles's *The Oxbow*

Cloaking worst disputes, masking deepest aches
until the haunting melts surely through,
snow discovers again the House of Solstice.
Always it has, always it will. Yet in this whiteness,
this place of prodigy, a centuries' old concerto
can dangle off the lobe of an ear
like a dusty aunt nudging another year
first, one sharp little twist then a second for cheer.
Or a friend whose words softly so crafted
seem to have been not written but transcend
into spring, like rainbow, like Virgo traversing the afternoon
of a Hudson River valley until they ascend
the green awe of the opposite ridge where self and space can mate.

2.

The Other Season

Out of this, in the barn, by the stall,
a daughter practices again her violin,
drawing a bow plucked from the tail of her horse
who will fly like Pegasus to lift her
over the double oxer and the slate
if that's what it takes
to keep her upright, keep her straight,

to free

    from an instrument

        music

where fingers confidently fret the neck,

independently find the notes,

a pitch

    just shy

        of perfect,

a timing

    just short

        of exact.

She skips onward with double stops;

Her mount perks in tune and nickers,

honed on the dustings of holiday,

her leather, her waxes, her string.

Let them romp in these presents and sing

knowing the sway of rhythm, even perfection,

cannot mind the gap or bridge the strait

between those who loved her first

and those whose love must wait.

Seventh Wave

Father sand dried underfoot.
Blue mother rolled in with her children;
she lined them up, lifted each in turn,
he caught and released
as she arced each toward shore.
In dull thuds they flopped
onto the beach and raced
to tag his vacuous toes,
then receded to their mother's embrace.
She gathered and laughed with watery arms,
lifted, and threw them toward him again.

On another beach, Nemo heard his mother sigh,
Every seventh wave is special.
Each will engulf your feet in laughter,
Then most shyly say goodbye.

Tea

A child once dreamed his life complete
when he could sit like his grandfather,
in velvet robe, sipping cosied tea,
and gaze at mist spun in from the sea.

This early March that child's daughter sleeps,
and he sits hushed by velvet within
the smooth-grained polish of a captain's chair.
He stares through greenhouse glass at dusts of snow,
and between the rising twists of steam sips his tea.

He listens to this house where dream matures to prayer:
"Where from here Grandad?" Then I was too young
to question beneath your calm. Then–there–
enough to sense peace outlasted your children's
consumed lungs and cancered breasts.

Then–you understood the orphan,
my brother's 'Dance on your grave' scream.
He knew only the triangle: Earth, anger, and pain.
Later, in your dirt, your garden centered by scent,
under your thatch, I learned what survival meant.

And now age sifts itself to a nuzzle of fur.
It seeps like leaves of tea that swell and steep
through clear, hot, liquid life infusing a soul with color.
I am your cup, bone-china thin but no rust,
no scald, no tremor will rim my thrust."

"Then rise," he said. "Measure your finest leaf
Into silver spoon and translucent porcelain cup.
It's time. Prepare great-granddaughter her tea."

Return to Thien Ngon, Vietnam

The road junction remains where it was, paved now.
Scooters battle trucks for right-of-way.
A village has evolved from the flattened landscape,

the arched, French bridge, blown then,
replaced by a concrete slab. The cauterized
bush regrown, lush in its early rainy season thirst.

This spot could have been an end—but wasn't,
like the tree in the forest which didn't
fall so wouldn't cry out for hearing.

An old priest remembered, his eye-contact
and imperceptible nod enough to fill decades.
Like daffodil bulbs after spring rain, shell casings still
turn up, tarnished reminder of the ugly beneath.
A323 gone, replaced by a national park for trees.
        For once someone did it right.

Acknowledgments of Previous Publication and Funding
Grateful thanks to these journals for first/previously publishing these pieces

"The Landing, Hancock Radio Telescope, *Hancock, New Hampshire*" was previously published as "Construction, Hancock Radio Telescope" in *Lungfish Review*, Vol. 1, No. 1, 1993, Portsmouth, NH, p. 43; and" in *Songs of Salamanders*, Hawk &Whippoorwill Press, 2020; *Adhoc Monadnock* e-zine, May, 2005; *Anemone*, Vol. 10, Winter Solstice Issue, 1992, Chester, VT, p. 23.

"In the Guise of the Canine," was first published as "Wolf", *Appalachia*, Winter 1984-1985, Vol. XLV, No.2, Appalachian Mtn. Club, Boston, MA, p. 57.

"Lyme Regis, before Nemo," *Tapestries*, Mt. Wachusett Community College, Gardner, MA, Fall 2006.

"Nemo's dinner in a Brattleboro Diner with Diana, daughter of a Nazi engineer," was first published as "World's Most Unlikely" in *Anemone*, Winter/Spring 1986, Vol. II, No.1, Chester, VT, p. 8.

"Prek Klok," *Smoky Quartz Quarterly*, Winter 2013, Monadnock Writers Group, Peterborough, NH; *The Monadnock Reader*, Monadnock Writers Group, p. 19, 1990. Peterborough, NH.

"Xóm Bào Dòn," *The Other Side of Sorrow: Poets Speak Out About Conflict, War and Peace*, Poetry Society of New Hampshire, 2006.

"Tunnel Rats," first published as "Lust," *The MacGuffin*, Vol. IV, No. 2, Schoolcraft College, 1987, p. 16.

"JD—FNG," *Poetry and the Vietnam Experience*. March, 1987, Canton, NY: St. Lawrence University.

"The Wound," The Monadnock Writer, 1985, p. 126, Monadnock Writers' Group, Peterborough, NH.

"War Dog Memorial," *The Black River: Death Poems* ,2024, p. 218; *The 2008 Poets' Guide to NH*, Poetry Society of New Hampshire, 2007, p. 11.

"In Pursuit of the Loon," *The 2008 Poets' Guide to NH*, Poetry Society of New Hampshire, 2007, p. 70; *The Loon Call*, Winter, 1990, p. 6; *Appalachia*, Vol. XLVII, No. 2, Dec. 15, 1988, Appalachian Mtn. Club, Boston, MA, p. 14.

"My Sea" previous version Untitled, *The Little Apple*, Summer 1982 No. 16, 1982, Worcester. MA, p. 16.

"Lonely Toes," *The Other Side of Sorrow, Poets Speak Out About Conflict, War and Peace,* Poetry Society of New Hampshire, 2006, p. 73; *Love and Trouble: An Anthology of Teachers' Writing,* Fall, 2002, Plymouth Writers Group, Plymouth, NH, p. 116.

"Upon Discovery of a Mammoth Hunter's Grave in France," originally published as The Mammoth Hunters, *The Issue,* No. 1, Autumn, 1998, Worcester County Poetry Association, Worcester, MA, p. 14; *Contemporary Foreign Literature,* No. 3, 1998, (China) trns, Zhang Ziqing, p. 141.

"Christmas and The Other Season," *Entelechy Intenational,* No. 4, p. 44-45, Fall 2006; *The Issue,* No. 2, Spring, 1999, Worcester County Poetry Association, Worcester, MA, p. 42.

"Tea," *The Monadnock Reader,* 1990, Monadnock Writers Group, Peterborough, NH, p. 176.

N.B. The First Edition of this book was published by Rodger Martin in 1990 with support from the New Hampshire State Council on the Arts.

This project was supported in part by funding from a Keene State College Adjunct Faculty Development Grant.

About the Author

Rodger Martin's *For All The Tea in Zhōngguó* (2019) follows *The Battlefield Guide*, (Hobblebush Books: 2010, 2013) and the selection of *The* Blue Moon *Series*, (Hobblebush Books: 2007) by *Small Press Review* which was one of its bi-monthly picks of the year. Most recently he has been a major contributor for NatureCulure®'s *Writing The Land®* anthologies.

Martin's distinctions include serving as a New Hampshire State Council on the Arts in Education roster artist and a touring artist for the New England States Touring Foundation, administered by the New England Foundation for the Arts. He has done artist-in-residency programs throughout New England. In 2012, he represented the United States as one of twelve poets participating in the City of Hangzhou's literary festival on West Lake, China. In 2015 he was a visiting poet at Nanjing University and Shanghai University of International Business and Economics, where in 2017 his poem "The Anchor" has been mounted at the reflecting pool where it was inspired. He returned to Yancheng with six other American poets as part of the *Poetry Bridging Continents III* Conference.

His awards include the 2024 Stanley Kunitz Medal for his lifelong commitment to poetry; an *Appalachia* award for poetry; a New Hampshire State Council on the Arts award for fiction; and fellowships from The National Endowment for the Humanities to study T.S. Eliot and Thomas Hardy at Oxford University and John Milton at Duquesne University.

In 2018, Meg Kearney chose one of Martin's poems for permanent trail mounting at Cathedral of The Pines. His publications include literary journals and anthologies throughout the United States and China where he also wrote a series of essays on American poetry for *The Yangtze River Journal*. He and six colleagues, as part of The Monadnock Pastoral Poets, have been featured in a new book *On the Monadnock: New Pastoral Poetry released in China in 2007.* 2013 marked the completion of a decade-long project with Dr. B. Eugene McCarthy to adapt all twelve books of the epic poem *Paradise Lost* for dramatic reading. His critical work "The Colonization of Paradise: Milton's Pandemonium and Montezuma's Tenochtitlan" published in *Comparative Literature Studies* broke new ground in Milton studies.

In addition to his writing, Martin teaches journalism and Creative Writing at Keene State College, co-advises *The Equinox*, the college's award-winning student news organization. He was managing editor of *The Worcester Review* for almost three decades, and for six years he directed New Hampshire's Poetry Foundation's Poetry Out Loud Project.

Rodger Martin was born in the amish country of Pennsylvania, lived in England as a child, and served as a combat engineer in Vietnam.

About NatureCulture Web

The mission of NatureCulture® is to help humans be in right relationship with the rest of the natural world. NatureCulture Web is our new imprint for books brought to us by like-minded authors and organizations.

Please see all NatureCulture's publications at:
https://www.nature-culture.net

Other NatureCulture® Books

2025
Dark Matter: Women Witnessing, Dreams Before Extinction, eds. Weil, Goslinga, Flyntz, & Bergeron

2024
The Black River: Death Poems
ed. Deirdre Pulgram-Arthen
Cayman Brac From Bluff to Sea
Writing the Land: The Connecticut River
Writing the Land: Wanderings I
Writing the Land: Wanderings II
Writing the Land: Virginia
Wriring the Land: Maine II, A Gathering

2023
Writing the Land: Youth Write the Land
Writing the Land: Currents
Writing the Land: Channels
Writing the Land: Streamlines
Migrations and Home: The Elements of Place, ed. Simon Wilson

From Root to Seed: Black, Brown, and Indigenous Poets Write the Northeast, ed. Samaa Abdurraqib

2022
Writing the Land: Foodways and Social Justice
Writing the Land: Windblown I
Writing the Land: Windblown II
Writing the Land: Maine
LandTrust, poems by Katherine Hagopian Berry

2021, 2024
Writing the Land: Northeast

Forthcoming (2025-2026)
Writing the Land: The Rensselaer Plateau
Writing the Land: Washington
Writing the Land: Doolin, Ireland
Writing the Land: The Cayman Islands
Writing the Land: Pathways
Writing the Land: The Great Forest of Aughty

www.ingramcontent.com/pod-product-compliance
Lightning Source LLC
Chambersburg PA
CBHW051228120626
46547CB00013B/1557